AQA
A Level Maths

AS Level

Authors
Paul Hunt, Steve Cavill,
Neil Tully, Rob Wagner

EXAM PRACTICE WORKBOOK

OXFORD
UNIVERSITY PRESS

Contents

About this book

This book contains three sets of write-in, mock exam papers for the AQA AS Level Maths exam (7356). Full details of this exam specification can be found on the AQA website.

http://web.aqa.org.uk/7356

There are two papers in each exam set: Paper 1 covers Pure and Mechanics and Paper 2 covers Pure and Statistics. Both papers are 90 minutes long and are each worth 80 marks.

The Large data set

The AS Level examination will assume that you are familiar with a Large data set (LDS). In the exam, some questions will be based on the LDS and may include extracts from it. It is AQA's intention that you should be taught using the LDS as this will give you a material advantage in the exam.

The LDS is a part of the data set originally used to produce the UK government's report 'Family Food Statistic 2014' (DEFRA, 2015). The LDS contains data on household purchases of food and drink between 2001 and 2014, subdivided by government office region. An Excel spreadsheet containing the LDS is available from the AQA website at the address given above.

Answers

The back of this book contains short answers to all the questions.
Full mark schemes for each mock paper can be found online.

https://global.oup.com/education/content/secondary/series/aqa-alevel-maths/aqaalevelmaths-answers

Formulae

In the exam, you will be provided with a 'Formulae for A Level Mathematics' booklet that is for use in AS Level and A Level Maths qualifications. These are provided at the end of this book. There are no statistical tables supplied for AS Level Maths.

Calculators

All papers are calculator papers. Since no statistical tables will be provided in the exam, you must make sure that you have a calculator and that you know how to use it, in particular to calculate probabilities for the binomial distribution. The rules on which calculators are allowed can be found in the Joint Council for General Qualifications document 'Instructions for conducting examinations' (ICE).

AS Level Mathematics
Paper 1 (Set A)

AQA

| Name | _____ | Class | _____ |
| Signature | _____ | Date | _____ |

Materials

You should have

- the booklet of formulae and statistical tables
- a graphical calculator.

Instructions

- Use a black pen for your working. Use a pencil for drawings.
- Answer **all** the questions in each of the two sections.
- Answer each question in the space provided for it; do **not** use the space provided for a different question. If you need extra space, ask for an additional answer booklet.
- All workings should be inside the box drawn around each page.
- To avoid losing method marks, show all necessary working.
- Include all rough working in this paper. If you do not want some work marked then cross it out.

Information

- Question marks are shown in square brackets.
- There is a maximum of 80 marks available for this paper.

Advice

- Unless asked for a proof, you may quote any of the formulae in the booklet.
- You may not need to use all the answer space provided.

Question	Mark
1	
2	
3	
4	
5	
6	
7	
8	
9	
10	
11	
12	
13	
Total	

Section A

Answer **all** questions in the spaces provided.

1 What is the solution set for the inequality $x^2 > 16$? **[1 mark]**

Circle your answer.

$x > 4$ $x < -4, x > 4$ $-4 < x < 4$ $x < -4$

2 a Integrate these expressions with respect to x

i x^3 **[1 mark]**

_____

_____

_____

ii x^{-2} **[1 mark]**

_____

_____

_____

iii $\dfrac{1}{x^5}$ **[1 mark]**

_____

_____

_____

b Hence, or otherwise, find $\displaystyle\int \left(x^2 + \frac{1}{x} \right)\left(\frac{3}{x^4} - 2x \right)\mathrm{d}x$ **[4 marks]**

_____

_____

_____

_____

_____

_____

_____

_____

_____

3 a   Use the factor theorem to show that $(x-6)$ and $(x+1)$ are both factors of the polynomial $f(x)$

$$f(x) = x^4 - x^3 - 22x^2 - 44x - 24$$

[3 marks]

b   Hence factorise $f(x)$ completely.

[4 marks]

4   In the diagram shown below, AB = AC = 10 cm, CE = 12 cm and DE = 5 cm.

Angle ABC = 75°

Find the ratio of the area of triangle ABC to the area of triangle ADE.

Express your answer in the form 1 : n **[5 marks]**

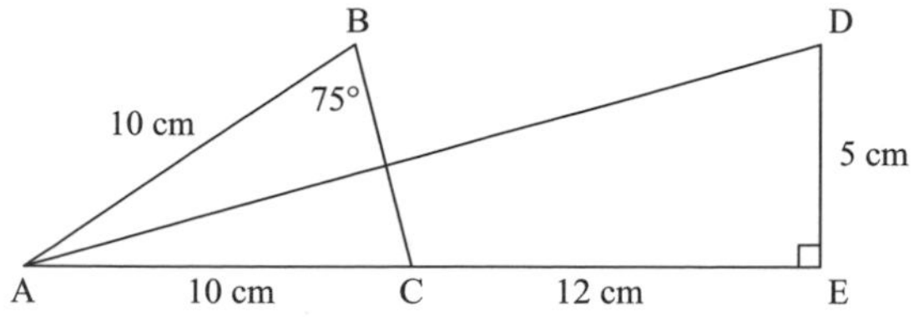

5 Find the exact solution(s) to this equation.

$$\sqrt{2x}\sqrt{x+2} = 3$$

[4 marks]

6   A circle C has equation $x^2 + y^2 - 6x - 12y + 41 = 0$

a   By completing the square, find the radius and coordinates of the centre of circle C.   **[4 marks]**

_____

_____

_____

_____

_____

_____

_____

_____

_____

_____

_____

_____

_____

_____

b   The circle C is a translation of the circle with equation $x^2 + y^2 = 4$

Describe this transformation geometrically.   **[2 marks]**

_____

_____

_____

_____

_____

_____

_____

_____

_____

_____

_____

_____

_____

_____

_____

6 c i The points A (1, 6) and B (3, 8) lie on circle C.

Find the equation of the perpendicular bisector of the line AB in the form $y = px + q$ **[6 marks]**

ii Find the exact coordinates of both points where the perpendicular bisector found in part **c i** intersects circle C.   **[6 marks]**

7 Solve these equations for all values of θ in the interval $0° \leq \theta \leq 90°$

a $\cos^2 \theta + \sin \theta = 1$ [4 marks]

b $\tan^2 5\theta = 1$ [3 marks]

8 A curve has equation $y = 3x^2 + \dfrac{3}{x^2}$

Find the area between the curve, the x-axis and the lines $y = a$ and $y = 3a$ where a is a positive constant. Give your answer in terms of a **[4 marks]**

End of section A

Section B

Answer **all** questions in the spaces provided.

9   Usain Bolt ran his fastest 100 m race in 9.58 s giving an average speed of 10.44 m s⁻¹.

What is his average speed in kilometres per hour?   **[1 mark]**

Circle your answer.

95.8   3.6   37.6   2.89

10   The velocity-time graph represents the motion of car.

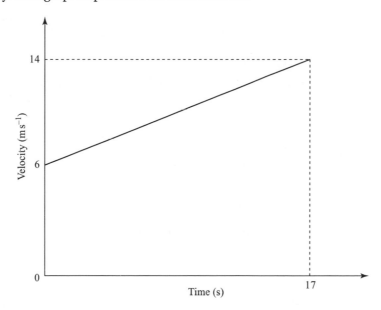

Calculate the distance travelled by the car during these 17 seconds.   **[2 marks]**

_____

_____

_____

_____

_____

_____

_____

_____

_____

_____

_____

_____

11   A force of magnitude 37 N has a component of 35 N in the x-direction.

a   Calculate the component of the force in the y-direction.   **[3 marks]**

b   Find the angle between the resultant force and the x-direction.   **[3 marks]**

12   In this question take g as 10 m s^{-2}

A particle is projected vertically downwards from point P with velocity 6 m s^{-1}

It reaches the ground after 3 s.

a   Calculate the height of point P above the ground.   **[2 marks]**

b   Calculate the speed of the particle at the instant it hits the ground.   **[2 marks]**

c   Calculate the time taken from the point of projection until the particle is 10 m above the ground.   **[5 marks]**

d   Explain how your answer to part **c** would change if you included the effect of air resistance on the particle.   **[2 marks]**

13   In this question take g as 10 m s^{-2}

Two masses of 5 kg and 8 kg are connected by a light, inextensible string that passes over a smooth, frictionless pulley. The masses are held stationary.

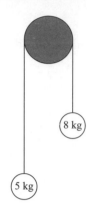

a   The masses are released from rest and accelerate at $a \text{ m s}^{-2}$
The tension in the string is T N.

By considering the forces acting on the 8 kg mass, show that $80 - T = 8a$ **[2 marks]**

_____

_____

_____

_____

_____

_____

_____

b   By first finding another equation, calculate both T and a **[5 marks]**

_____

_____

_____

_____

_____

_____

_____

_____

_____

_____

_____

_____

_____

End of questions

AS Level Mathematics
Paper 2 (Set A)

AQA

| Name | | Class | |
| Signature | | Date | |

Materials

You should have

- the booklet of formulae and statistical tables
- a graphical calculator.

Instructions

- Use a black pen for your working. Use a pencil for drawings.
- Answer **all** the questions in each of the two sections.
- Answer each question in the space provided for it; do **not** use the space provided for a different question. If you need extra space, ask for an additional answer booklet.
- All workings should be inside the box drawn around each page.
- To avoid losing method marks, show all necessary working.
- Include all rough working in this paper. If you do not want some work marked then cross it out.

Information

- Question marks are shown in square brackets.
- There is a maximum of 80 marks available for this paper.

Advice

- Unless asked for a proof, you may quote any of the formulae in the booklet.
- You may not need to use all the answer space provided.

Question	Mark
1	
2	
3	
4	
5	
6	
7	
8	
9	
10	
11	
12	
13	
Total	

Section A

Answer **all** questions in the spaces provided.

1   Which of these expressions is equivalent to $\dfrac{1}{5-2\sqrt{3}}$?   **[1 mark]**

Circle your answer.

$$5+2\sqrt{3} \qquad \frac{5-2\sqrt{3}}{37} \qquad \frac{5+2\sqrt{3}}{13} \qquad \frac{5-2\sqrt{3}}{13}$$

2   What is $\dfrac{\mathrm{d}}{\mathrm{d}x}(\pi^2)$?   **[1 mark]**

Circle your answer.

$$2\pi \qquad 0 \qquad \pi^2 \qquad 1$$

3   Two numbers are in the ratio $6:11$
When 9 is subtracted from each number, their ratio changes to $3:8$

What are the two numbers?   **[3 marks]**

4 Write this expression as a single logarithm with a coefficient of 1

$$\frac{1}{2}\log_{10} b - \log_{10} 1 - \frac{1}{4}\log_{10} c^2$$

[4 marks]

5 What is the coefficient of x^{10} in the binomial expansion of $\left(x^2 - \frac{1}{4} \right)^8$?

[2 marks]

6 Use differentiation from first principles to find the derivative of this function with respect to x

$$q(x) = 4x^2 - 2x$$

[5 marks]

7 It is claimed that the data in the table is consistent with a relationship of the form $y = ab^x$

x	1	5	7	8	12
y	1.67	2.19	2.51	2.68	3.51
$\log_{10} y$					

a i Show that this relationship can also be written in the linear form

$\log_{10} y = x \log_{10} b + \log_{10} a$ **[3 marks]**

_____

_____

_____

_____

_____

_____

_____

ii By completing the table above and then drawing a suitable graph, explain whether or not this claim is true. **[3 marks]**

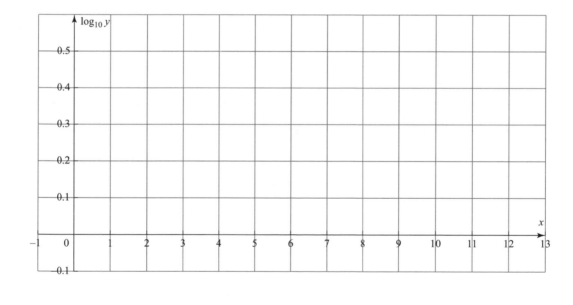

_____

_____

_____

_____

_____

_____

_____

_____

7 b Estimate the values of the constants a and b
Give your answers to 2 significant figures.   **[5 marks]**

Estimate the values of the constants a and b
Give your answers to 2 significant figures.   **[5 marks]**

8   The circles C_1 and C_2 share a common chord AB.

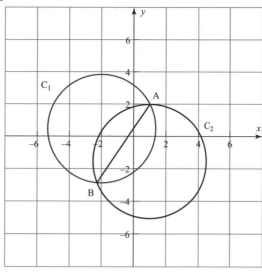

The equations of the circles are

$$C_1 : \ 4(x-1)^2 + 4\left(y+\frac{3}{2}\right)^2 = 49$$

$$C_2 : \ 4(x+2)^2 + 4\left(y-\frac{1}{2}\right)^2 = 45$$

a   **i**   Find the equation of the chord AB in the form $y = ax + b$ **[4 marks]**

8 a ii Hence prove that the coordinates of point A are (1, 2) and point B are $\left(-\dfrac{29}{13}, -\dfrac{37}{13}\right)$ **[5 marks]**

_____

_____

_____

_____

_____

_____

_____

_____

_____

_____

_____

_____

_____

_____

_____

_____

_____

b Find the distance between the points A and B.
Give your answer to 3 decimal places.   **[2 marks]**

_____

_____

_____

_____

_____

_____

_____

_____

_____

_____

_____

_____

_____

_____

_____

_____

_____

9   A cuboid has a base of length $4x$ cm and width $3x$ cm.
The combined total of all edges of the cuboid is 16 cm.

a   Find an expression for the height of the cuboid in terms of x **[2 marks]**

b   Show that the surface area of the cuboid is $56x - 74x^2$ **[3 marks]**

9 c i Find the exact value of the maximum possible surface area of the cuboid. **[7 marks]**

ii State the exact dimensions of the cuboid when the surface area is at its maximum. **[3 marks]**

End of section A

10   Coloured sweets are produced in red, green and blue varieties. A large number are placed in a bowl and one is picked out at random without looking. These two events cannot happen at the same time.

'A red sweet is picked'

'A blue sweet is picked'

a   What term best describes these events?   **[1 mark]**

Circle your answer.

Independent   Mutually exclusive   Random   Conditional

b   The bowl contains 36 red, 48 green, and 16 blue sweets.
State the probability of picking a red sweet if one sweet is picked.   **[1 mark]**

_____

_____

_____

_____

_____

_____

11   As part of the Living Costs and Food Survey, data is collected on the average amount of various foodstuffs eaten per person per week by people in the UK. In 2014, people ate an average of 111 g of cheese per person per week. A newspaper takes a sample from a local town in 2015 to try to identify any changes in the amount of cheese eaten per person per week.

a   The newspaper only takes a sample from its own town, claiming that it doesn't expect habits in that town to differ from those of the whole UK population and so it doesn't need to take a sample from other locations.

Give the name of this sampling method.   **[1 mark]**

_____

_____

_____

_____

_____

_____

11 b The newspaper's sample data are grouped in this table showing the average mass of cheese eaten per person per week to the nearest 10 g and the number of people who eat that amount.
There are 44 people in the sample.

Mass of cheese (g)	≤ 60	70	80	90	100	110	120	130	140	150	≥ 160
Number of people	5	3	4	10	7	3	5	4	1	2	0

Estimate the probability that a randomly-chosen person in this sample eats *at least* an average of 115 g of cheese per week.

[2 marks]

c Data is also collected on the amount of meat the same people eat on average per week.
It is found that the amount of meat eaten is independent of the amount of cheese eaten.

Explain what it means for these quantities to be independent.

[1 mark]

12   The percentage scores that each student in a class achieves on their Maths and English exams are shown in the scatter diagram.

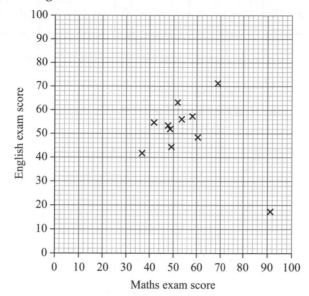

a   Circle any outlier(s) in the data, and explain why you believe they are outliers.   **[2 marks]**

_____

_____

_____

_____

_____

_____

_____

_____

_____

b   Ignoring the outliers you identified in part **a**, add a line of best fit to the scatter graph.   **[1 mark]**

c   One student scored 65% on the Maths exam but was ill for the English exam, so their data has been omitted from the scatter diagram.

Using your line of best fit, estimate the score they might have achieved on the English exam. Show how you determined your estimate.   **[2 marks]**

_____

_____

_____

_____

_____

_____

_____

_____

12 d Using the data from the scatter diagram, fill in this grouped frequency table. [3 marks]

Maths exam score	$0 \leq x < 40$	$40 \leq x < 45$	$45 \leq x < 50$	$50 \leq x < 55$	$55 \leq x < 60$	$60 \leq x < 70$	$70 \leq x < 100$
Frequency							

e A student is randomly chosen from the class.

i Calculate the probability that they scored at least 45% but less than 60%. [2 marks]

ii Calculate the probability that they scored at least 50% given that they score at least 45% but less than 60%. [3 marks]

13 In a ready meal factory there is a fixed probability of 6% that any one packet isn't fit for sale, independent of any other packet. The factory manager believes that the fixed probability of a packet not being fit for sale has increased. To test this hypothesis the manager wants to take a sample of packets produced in the factory during one week.

a i The factory manager wants to use simple random sampling to take a sample of size 18 from the 3000 packets produced that week.

Describe how the manager can take a simple random sample. [3 marks]

13 a ii Explain why this sample is not expected to be biased. [1 mark]

_____

_____

_____

_____

_____

b In the sample of 18 packets, three of them are not fit for sale.

Carry out a hypothesis test at the 5% significance level to investigate if the manager's belief is supported by the evidence. [4 marks]

_____

_____

_____

_____

_____

_____

_____

_____

_____

_____

_____

_____

_____

_____

_____

_____

_____

_____

_____

_____

End of questions

AS Level Mathematics
Paper 1 (Set B)

AQA

Name _____ Class _____

Signature _____ Date _____

Materials

You should have

- the booklet of formulae and statistical tables
- a graphical calculator.

Instructions

- Use a black pen for your working. Use a pencil for drawings.
- Answer **all** the questions in each of the two sections.
- Answer each question in the space provided for it; do **not** use the space provided for a different question.
  If you need extra space, ask for an additional answer booklet.
- All workings should be inside the box drawn around each page.
- To avoid losing method marks, show all necessary working.
- Include all rough working in this paper. If you do not want some work marked then cross it out.

Information

- Question marks are shown in square brackets.
- There is a maximum of 80 marks available for this paper.

Advice

- Unless asked for a proof, you may quote any of the formulae in the booklet.
- You may not need to use all the answer space provided.

Question	Mark
1	
2	
3	
4	
5	
6	
7	
8	
9	
10	
11	
12	
13	
14	
15	
Total	

Section A

Answer **all** questions in the spaces provided.

1   Which of these expressions is the same as $e^x \times e^x$?   **[1 mark]**

Circle your answer(s).

$$e^{x^2} \qquad 2e^x \qquad e^{2x} \qquad (e^x)^2$$

2   What is the period of the graph of the function $y = \sin 2\theta$?   **[1 mark]**

Circle your answer.

$$90° \qquad 180° \qquad 360° \qquad 720°$$

3   Use the fact that $\log_{10} 3 = 0.47712$ to 5 decimal places to calculate the following to 4 decimal places.

You must show your working.

a $\log_{10} 27$ **[3 marks]**

_____

_____

_____

_____

_____

_____

b $\log_{10} 300$ **[3 marks]**

_____

_____

_____

_____

_____

_____

_____

_____

3 **c** $\log_{10}\left(\dfrac{300}{27}\right)$ **[2 marks]**

_____

_____

_____

_____

_____

_____

_____

_____

_____

4 Sketch each of the following graphs on the separate sets of axes provided.

Make sure that you clearly state the equations of any asymptotes, and the coordinates of all points where the graphs intersect the coordinate axes.

a $y - 2 = (x + 3)^2$ **[3 marks]**

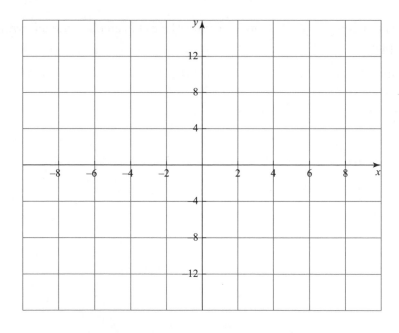

_____

_____

_____

_____

_____

_____

_____

4 **b** $y = (4-x)(2x-3)^2$ **[3 marks]**

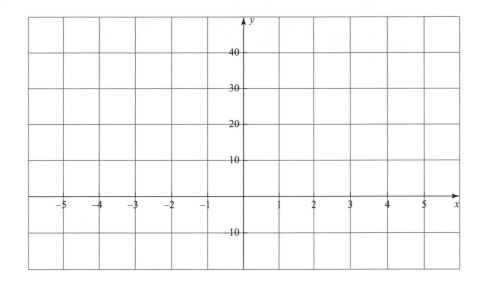

_____

_____

_____

_____

_____

_____

c $y = \dfrac{1}{2(x+5)^2}$ **[3 marks]**

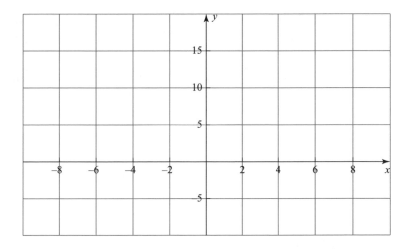

_____

_____

_____

_____

_____

_____

4   d $y = 2\sin 2x$ **[3 marks]**

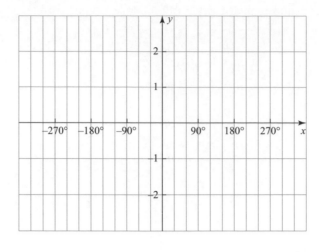

_____

_____

_____

_____

_____

_____

_____

_____

5   Prove that $t^2 - t$ is even for all positive integers, t **[4 marks]**

_____

_____

_____

_____

_____

_____

_____

_____

_____

_____

_____

_____

_____

6 A curve C and a line L have equations $y = 6x^2 + 8x - 4$ and $y = p(2x - 1)$

a Show that the x-coordinate of any point of intersection of C and L satisfies the equation

$$6x^2 + 2(4 - p)x + p - 4 = 0$$

[1 mark]

b C and L intersect at two distinct points.

i Show that $p^2 - 14p + 40 > 0$

[3 marks]

ii Find the possible values of p

[4 marks]

7   Two sides of a parallelogram are 6.6 cm and 8.9 cm long.

One of the diagonals of the parallelogram is 5.7 cm long.

What is the area of the parallelogram?

Give your answer to one decimal place.   **[4 marks]**

8 a Write down the first three terms, in ascending powers of x, of the binomial expansion of $(2 - x)^8$. Write each term in its simplest form. **[3 marks]**

b Use your answer to part **a** to estimate the value of 1.99^8 **[1 mark]**

9 A biologist is growing bacteria in a dish and has successfully modelled the population, p, at time t (hours) after the start of the experiment by the function $p = 24e^{2t}$

a An assistant measures the rate of increase in the population to be 1200 bacteria/hour. Give an exact expression for the time this measurement was made. **[4 marks]**

b i Give one criticism of the biologist's model. [1 mark]

_____

_____

_____

_____

_____

_____

_____

ii Describe how the model could be improved. [1 mark]

_____

_____

_____

_____

10 Prove that the function $4x^3 - 6x^2 + 6x - 1$ is an increasing function [5 marks]

_____

_____

_____

_____

_____

_____

_____

_____

_____

_____

_____

_____

_____

_____

_____

_____

_____

End of section A

Section B

Answer **all** questions in the spaces provided.

11   All SI units in mechanics can be written using the kilogram (kg), the metre (m) and the second (s). For example, velocity can be written as m/s or m s^{-1}

Write the unit of force, the newton, using just kg, m and s.   **[1 mark]**

Circle your answer.

$$kg\,m \qquad kg\,m\,s^{-1} \qquad kg\,m^2\,s^{-1} \qquad kg\,m\,s \qquad kg\,m\,s^{-2}$$

12   The three points A, B and C have positions vectors

$$\overrightarrow{OA} = \mathbf{i} + \mathbf{j}, \quad \overrightarrow{OB} = 3\mathbf{i} - 2\mathbf{j}, \quad \overrightarrow{OC} = 7\mathbf{i} - 8\mathbf{j}$$

Prove that A, B and C are collinear.   **[6 marks]**

13   This displacement-time graph models the motion of a bicycle.

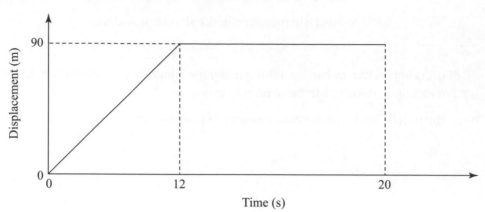

a   Describe what the bicycle is doing from $t = 12$ s to $t = 20$ s.   **[1 mark]**

_____

_____

_____

b   Calculate the velocity of the bicycle during the first 12 seconds.   **[2 marks]**

_____

_____

_____

_____

_____

_____

_____

_____

14   A particle moves such that its velocity, v m s^{-1}, after time, t s, is given by
$v = 12t - 3t^2$

a   Find an expression for the acceleration of the particle after t seconds.   **[2 marks]**

_____

_____

_____

_____

_____

_____

_____

14 b   The initial displacement of the particle is 5 m.

Show that the displacement when $t = 4$ s is 37 m.   **[5 marks]**

_____

_____

_____

_____

_____

_____

_____

_____

_____

_____

_____

_____

_____

c   Find the times when the particle is at rest.   **[3 marks]**

_____

_____

_____

_____

_____

_____

_____

_____

_____

_____

_____

_____

_____

_____

15   A car of mass 800 kg is pulling a trailer of mass 200 kg.

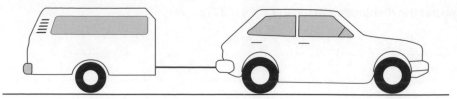

The car and trailer experience resistances of 100 N and 250 N respectively.
The car exerts a driving force of 500 N.

a   Find the acceleration of the car and trailer.   **[4 marks]**

_____

_____

_____

_____

_____

_____

_____

_____

b   Find the tension in the tow bar.   **[3 marks]**

_____

_____

_____

_____

_____

_____

_____

_____

_____

_____

_____

_____

End of questions

Name		Class	
Signature		Date	

Materials

You should have

- the booklet of formulae and statistical tables
- a graphical calculator.

Instructions

- Use a black pen for your working. Use a pencil for drawings.
- Answer **all** the questions in each of the two sections.
- Answer each question in the space provided for it; do **not** use the space provided for a different question. If you need extra space, ask for an additional answer booklet.
- All workings should be inside the box drawn around each page.
- To avoid losing method marks, show all necessary working.
- Include all rough working in this paper. If you do not want some work marked then cross it out.

Information

- Question marks are shown in square brackets.
- There is a maximum of 80 marks available for this paper.

Advice

- Unless asked for a proof, you may quote any of the formulae in the booklet.
- You may not need to use all the answer space provided.

Question	Mark
1	
2	
3	
4	
5	
6	
7	
8	
9	
10	
11	
12	
13	
Total	

Section A

Answer **all** questions in the spaces provided.

1 Which of these expressions is the same as $\frac{\frac{1}{x}}{x}$? [1 mark]

Circle your answer(s).

$$\frac{1}{x^2} \qquad 1 \qquad \frac{1}{2x} \qquad x^2$$

2 What is the gradient of the straight line whose equation is $2x - 3y - 6 = 0$? [1 mark]

Circle your answer.

$$-\frac{3}{2} \qquad 2 \qquad \frac{2}{3} \qquad -\frac{2}{3}$$

3 James says he has solved the equation below but Linda claims that his solution is wrong.

Who is correct? Explain your answer.

$$x^2 - 6x + 8 = 3$$
$$(x-2)(x-4) = 3$$
$$x - 2 = 3 \text{ or } x - 4 = 3$$
$$x = 5 \text{ or } x = 7 \qquad \text{[1 mark]}$$

_____

_____

_____

_____

_____

_____

_____

_____

_____

_____

_____

_____

_____

_____

4   Solve this pair of simultaneous equations.

$$2y^2 - x^2 + xy = 14$$

$$5x + 4y = 7$$

[6 marks]

5 **a**   Solve this equation, to 1 decimal place, for values of θ in the interval $0 < \theta < 360°$

$$4\cos^2\theta + \sin\theta = 2$$

[5 marks]

b   Hence solve this equation, to 1 decimal place, for values of x in the interval $0 < x < 180°$

$$4\cos^2(2x + 30°) + \sin(2x + 30°) = 2$$

[2 marks]

6 The value, V, of a particular mobile phone depends on t, the age of the phone in years from the date of purchase, and is given by the formula below.

$$V = 150\left(1 + 4e^{-\frac{t}{3}}\right)$$

a What was the purchase price of the phone? **[1 mark]**

b How much will the phone be worth two years after it was purchased? **[1 mark]**

c How long after it was first purchased will the phone have lost half of its initial value? **[4 marks]**

6 d   Sketch a graph of V against t, showing how the value of the phone varies over time.   **[4 marks]**

V

t

_____

_____

_____

_____

_____

_____

_____

_____

_____

_____

_____

_____

_____

_____

_____

_____

e   Do you think the model is realistic? Explain your answer.   **[1 mark]**

_____

_____

_____

_____

_____

_____

_____

_____

7   The diagram shows a kite, whose diagonals are of length a cm and $\left(4+\sqrt{3}\right)$ cm.

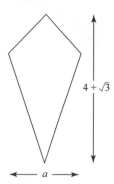

$4+\sqrt{3}$

$\longleftarrow a \longrightarrow$

The area of the kite is $\left(\dfrac{5}{2}-\sqrt{3}\right)$ cm^2.

Calculate the *exact* length a. You must show your workings.   **[5 marks]**

8   The graph below shows a quadratic curve intersecting the x-axis at the points $(a, 0)$ and $(1, 0)$ and the y-axis at the point $(0, 1)$

The graph has its vertex at the point $\left(\dfrac{1}{4}, \dfrac{9}{8}\right)$

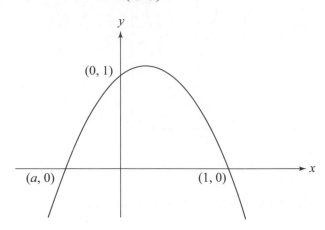

a   Find the equation of the quadratic curve.   **[7 marks]**

_____

_____

_____

_____

_____

_____

_____

_____

_____

_____

_____

_____

_____

_____

_____

_____

_____

_____

_____

_____

_____

_____

8 b By factorising the equation found in part **a**, find the value of a [2 marks]

c Find the area bounded by the curve and the x-axis.

You must show your workings. [5 marks]

9 The second derivative of a function, $u(x)$, is given by $\dfrac{d^2u}{dx^2} = 30x + 8$

When $x = -1$, $\dfrac{du}{dx} = 7$

a Find an expression for $\dfrac{du}{dx}$ in terms of x [4 marks]

b When $x = -1$, $u = -2$

Find an expression for u in terms of x [3 marks]

End of section A

Section B

Answer **all** questions in the spaces provided.

10   A hypothesis test is being performed.

The probability of obtaining a result in the critical region is 3.8%.

a   Circle the most likely significance level for the test.   **[1 mark]**

$$1\% \qquad 2\% \qquad 5\% \qquad 10\%$$

b   The critical region is $X \geq 37$. State any critical value(s).   **[1 mark]**

_____

_____

_____

_____

11   The probability that an apple from a multipack weighs under 80 g is 0.15

The probability that an apple from a multipack is bruised is 0.05

The probability of being underweight and the probability of being bruised are independent.

a   Find the probability that a randomly-chosen apple both weighs under 80 g and is bruised.   **[1 mark]**

_____

_____

_____

_____

11 b Two apples are chosen from *different* multipacks.

    i Find the probability that exactly one weighs under 80 g.   **[1 mark]**

_____

_____

_____

_____

    ii Find the probability that at least one weighs under 80 g and at least one is bruised.   **[2 marks]**

_____

_____

_____

_____

12 Data is collected, using opportunity sampling, and the following results are obtained.

    2.3   2.6   2.7   3.5   3.8   4.1   4.2   4.5   5.5   6.8

    a **i** Explain what is meant by opportunity sampling.   **[1 mark]**

_____

_____

_____

_____

    ii Calculate the mean and standard deviation of this sample.   **[2 marks]**

_____

_____

_____

_____

_____

12 b Another sample is taken using systematic sampling.
That sample has mean 3.2 and standard deviation 0.7

   i Give one reason why opportunity sampling is more likely than systematic sampling to provide a biased sample. **[1 mark]**

_____

_____

_____

_____

   ii Compare and contrast the two samples using the information given. **[2 marks]**

_____

_____

_____

_____

_____

13 The volume of ice cream sold per day by a grocer is monitored over time and presented in this histogram.

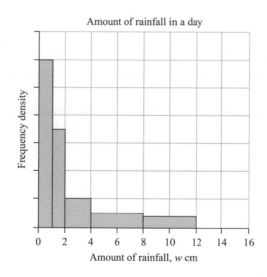

Amount of rainfall in a day

Frequency density

Amount of rainfall, w cm

a Use the histogram to complete this probability distribution table. **[3 marks]**

Volume of ice cream, w litres	$0 \leq w < 1$	$1 \leq w < 2$	$2 \leq w < 4$	$4 \leq w < 8$	$8 \leq w < 12$	$12 \leq w$
Frequency	12					

13 b   Use your table in part **a** to estimate the probability that at least six litres of ice cream are sold on a random day.   **[2 marks]**

_____
_____
_____
_____
_____
_____
_____
_____
_____
_____
_____
_____
_____
_____
_____

c   The grocer thinks that the rain forecast over the next two weeks is likely to reduce ice cream sales. They look to see if there are significantly fewer days with sales of six litres or more of ice cream.

i   State two conditions required for a binomial distribution to be an appropriate model and explain why they apply to the grocer's problem.   **[4 marks]**

_____
_____
_____
_____
_____
_____
_____
_____
_____
_____
_____
_____
_____
_____

13 c ii Over the two rainy weeks there were no days on which six litres or more of ice cream were sold.

By calculating the p-value of the result, perform a hypothesis test at the 10% level to see if there is less ice cream being sold during the rainy weeks. State your hypotheses clearly.

[6 marks]

_____

_____

_____

_____

_____

_____

_____

_____

_____

_____

_____

_____

_____

_____

_____

_____

_____

_____

_____

_____

_____

_____

_____

_____

_____

End of questions

AS Level Mathematics
Paper 1 (Set C)

AQA

| Name | | Class | |

| Signature | | Date | |

Materials

You should have

- the booklet of formulae and statistical tables
- a graphical calculator.

Instructions

- Use a black pen for your working.
  Use a pencil for drawings.
- Answer **all** the questions in each of the two sections.
- Answer each question in the space provided for it; do **not** use the space provided for a different question.
  If you need extra space, ask for an additional answer booklet.
- All workings should be inside the box drawn around each page.
- To avoid losing method marks, show all necessary working.
- Include all rough working in this paper.
  If you do not want some work marked then cross it out.

Question	Mark
1	
2	
3	
4	
5	
6	
7	
8	
9	
10	
11	
12	
Total	

Information

- Question marks are shown in square brackets.
- There is a maximum of 80 marks available for this paper.

Advice

- Unless asked for a proof, you may quote any of the formulae in the booklet.
- You may not need to use all the answer space provided.

Section A

Answer **all** questions in the spaces provided.

1   The area of a rectangle must remain fixed.

However, the length, L, and the width, W, are allowed to vary.

a   Which of the following expression(s) is/are true? Here k is a constant.   **[1 mark]**

Circle your answer(s).

$$L \propto W \qquad\qquad L \propto \frac{1}{W} \qquad\qquad L = kW \qquad\qquad L = \frac{k}{W}$$

b   Which of these graphs best illustrates the relationship between L and W?   **[1 mark]**

Circle your answer.

A
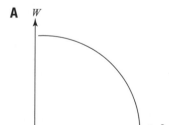

B
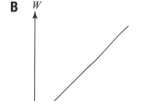

C
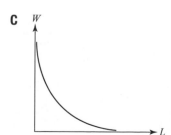

D
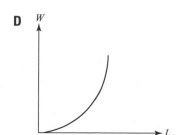

2 a   There are no prime numbers between 115 and 125 inclusive.

Is this statement correct?

Prove or disprove your answer.   **[3 marks]**

_____

_____

_____

_____

_____

_____

_____

_____

_____

_____

_____

_____

_____

b   If the product ab is a rational number, then the numbers a and b are also rational numbers.

Disprove this statement.   **[1 mark]**

_____

_____

_____

_____

_____

_____

_____

_____

_____

_____

_____

_____

_____

_____

_____

_____

3 **a**   A point on the unit circle has coordinates $\left(-\dfrac{\sqrt{2}}{2}, \dfrac{\sqrt{2}}{2}\right)$

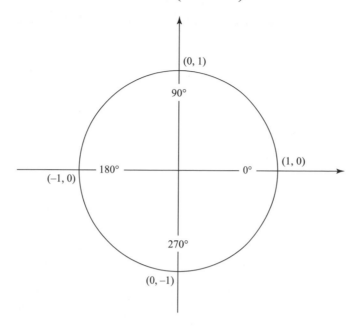

What obtuse angle does the point represent?   **[1 mark]**

_____

_____

_____

_____

b $\cos\alpha = \dfrac{15}{17}$, where α is a reflex angle.

i   Derive an exact value for $\sin\alpha$ **[4 marks]**

_____

_____

_____

_____

_____

_____

_____

_____

_____

_____

_____

_____

_____

3 b ii Derive an exact value for $\tan \alpha$

[2 marks]

c Prove this trigonometric identity.

$$\frac{\cos x}{1-\sin x} \equiv \frac{1+\sin x}{\cos x}$$

[4 marks]

4 **a** Solve these inequalities exactly.

i
$$-2 \leq \frac{6-3x}{5} < 6$$
[3 marks]

ii
$$4(2-x)-3(8-3x) < 4-3(2x+1)+x$$
[3 marks]

4 b Habib's teacher has challenged him to solve the inequality $\dfrac{x-6}{x+4} < 6$

Habib's attempt is shown below.

$$\frac{x-6}{x+4} < 6$$

$$x - 6 < 6(x + 4)$$

$$x - 6 < 6x + 24$$

$$5x > -30$$

$$x > -6$$

Is Habib's solution correct?

Justify your answer. **[2 marks]**

_____

_____

_____

_____

_____

_____

_____

_____

5 a It is claimed that the two variables, x and y, are connected by a relationship of the form $y = kx^n$ where k and n are constants.

    i Show that this relationship can also be written in the linear form

$$\log_{10} y = \log_{10} k + n \log_{10} x$$ **[3 marks]**

_____

_____

_____

_____

_____

_____

_____

_____

_____

_____

_____

5 a ii By completing the table of x and y values and then drawing a suitable straight-line graph, test whether or not this claim is true. **[4 marks]**

x	3	4	6	9	10
y	14.50	22.97	43.95	84.09	99.53
$\log_{10}x$					
$\log_{10}y$					

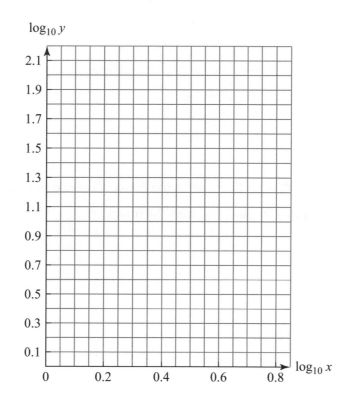

_____

_____

_____

b Estimate the values of the constants k and n, giving your answers to 1 decimal place. **[4 marks]**

_____

_____

_____

_____

_____

_____

_____

_____

_____

_____

6   The graph shows the gradient function, $y = f'(x)$, of a function $y = f(x)$

a   On the same axes, sketch the graph of $y = f''(x)$ [3 marks]

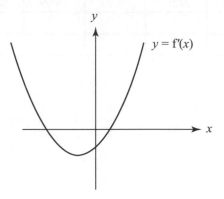

b   On this copy of the graph of $y = f'(x)$, sketch the graph of the original function $y = f(x)$

You may assume that $y = f(x)$ passes through the origin.   [3 marks]

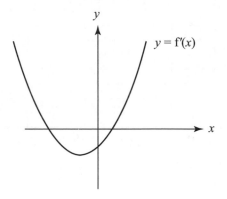

7 a   Three points, A, B and C, have position vectors $(2\mathbf{i} + 3\mathbf{j})$, $(6\mathbf{i} - 4\mathbf{j})$ and $(-5\mathbf{i} - 3\mathbf{j})$ respectively.

What is the perimeter of the triangle ABC?

Give your answer to 3 significant figures.   [5 marks]

_____

_____

_____

_____

_____

_____

_____

_____

_____

_____

_____

7 b In the diagram, $\overrightarrow{OD} = \mathbf{d}$, $\overrightarrow{OE} = \mathbf{e}$ and $\overrightarrow{OF} = \mathbf{f}$

The midpoints of $\overrightarrow{EF}$, $\overrightarrow{FD}$ and $\overrightarrow{DE}$ are G, H and J respectively.

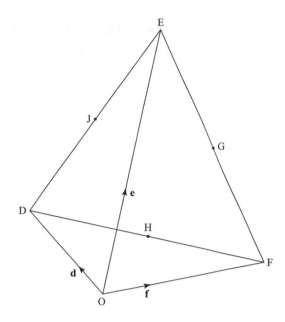

i Find, in terms of $\mathbf{e}$, $\mathbf{f}$ and $\mathbf{g}$, the sum $\overrightarrow{OG} + \overrightarrow{OH} + \overrightarrow{OJ}$ **[5 marks]**

_____

_____

_____

_____

_____

_____

_____

_____

_____

_____

_____

ii Thus write down $\overrightarrow{OG} + \overrightarrow{OH} + \overrightarrow{OJ}$ in terms of $\overrightarrow{OD}$, $\overrightarrow{OE}$ and $\overrightarrow{OF}$ **[1 mark]**

_____

_____

_____

_____

End of section A

Section B

Answer **all** questions in the spaces provided.

8   A canoeist can paddle in still water at 4 km h^{-1}. She wants to paddle straight across a river to the opposite bank. The water is flowing with a current of 3 km h^{-1}

At what angle to the riverbank must she paddle?

Circle your answer.   **[1 mark]**

$$36.9° \qquad 41.4° \qquad 45° \qquad 48.6°$$

9   A particle of mass 3 kg is acted on by forces of 5**i** N and 7**j** N.

Calculate the acceleration of the particle.   **[3 marks]**

_____

_____

_____

_____

_____

_____

_____

_____

_____

_____

_____

_____

_____

_____

_____

_____

_____

10 a When a skydiver jumps from an aircraft he accelerates until he reaches his *terminal velocity*.
The terminal velocity of a skydiver is 192 km h^{-1}

Convert 192 km h^{-1} into m s^{-1} [1 mark]

_____

_____

_____

_____

_____

_____

b Jo says that she should **not** use the formula $v = u + at$ to calculate the time it takes the skydiver to reach terminal velocity.

Explain why Jo is correct. [1 mark]

_____

_____

_____

_____

_____

_____

10 c When a skydiver reaches terminal velocity he stops accelerating.

Describe the forces acting on a skydiver falling at terminal velocity and calculate the air resistance acting on a skydiver of mass 80 kg.

You should use $g = 9.81 \text{ m s}^{-2}$ **[3 marks]**

_____

_____

_____

_____

_____

_____

_____

_____

_____

11 This velocity-time graph represents the motion of a particle moving along a straight line.

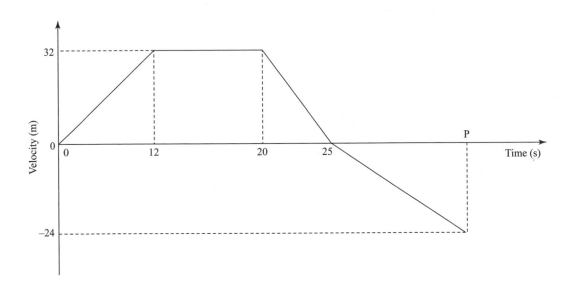

a Calculate the acceleration of the particle during the first 12 seconds.   **[2 marks]**

_____

_____

_____

_____

_____

_____

11 b When is the particle at rest? **[1 mark]**

_____

_____

_____

_____

c Given that the particle ends up in the same place that it started, find P. **[5 marks]**

_____

_____

_____

_____

_____

_____

_____

_____

_____

_____

12 A particle starts from the origin with initial velocity $3 \, \mathrm{m\,s^{-1}}$ and moves such that its acceleration after time t s is given by $a = 6t - 4$

a Find a formula for the velocity, $v \, \mathrm{m\,s^{-1}}$, of the particle after t seconds. **[3 marks]**

_____

_____

_____

_____

_____

_____

_____

_____

_____

_____

_____

_____

12 b Show that the particle never returns to the origin. **[7 marks]**

End of questions

Name		Class	
Signature		Date	

Materials

You should have

- the booklet of formulae and statistical tables
- a graphical calculator.

Instructions

- Use a black pen for your working.
  Use a pencil for drawings.
- Answer **all** the questions in each of the two sections.
- Answer each question in the space provided for it; do **not** use the space provided for a different question.
  If you need extra space, ask for an additional answer booklet.
- All workings should be inside the box drawn around each page.
- To avoid losing method marks, show all necessary working.
- Include all rough working in this paper.
  If you do not want some work marked then cross it out.

Question	Mark
1	
2	
3	
4	
5	
6	
7	
8	
9	
10	
11	
12	
Total	

Information

- Question marks are shown in square brackets.
- There is a maximum of 80 marks available for this paper.

Advice

- Unless asked for a proof, you may quote any of the formulae in the booklet.
- You may not need to use all the answer space provided.

Section A

Answer **all** questions in the spaces provided.

1   Which of these expressions is the same as $\sqrt{8}$?   **[1 mark]**

Circle your answer(s).

$$4\sqrt{2} \qquad 2\sqrt{2} \qquad \pm 4\sqrt{2} \qquad \pm 2\sqrt{2}$$

2   A graph shows $y = 2\sin 3x$

What is the amplitude of the graph?   **[1 mark]**

Circle your answer(s).

$$60 \qquad 120 \qquad 2 \qquad 4$$

3   Daryl has solved the equation $3\sin^2 x = \sin x$, where $0° \leq x \leq 180°$

His solution is shown below.

$$3\sin^2 x = \sin x$$
$$3\sin x = 1$$
$$\sin x = \frac{1}{3}$$
$$x = 19.5°, \ x = 160.5° \ (1 \text{ dp})$$

a   Why is Daryl's solution incorrect?   **[1 mark]**

_____

_____

_____

_____

_____

_____

_____

_____

_____

_____

b   Solve the equation correctly.   **[3 marks]**

_____

_____

_____

_____

_____

_____

_____

_____

_____

_____

_____

_____

_____

_____

4   Two consecutive, positive, even numbers are chosen.

The square of both numbers is taken and then the smaller answer is multiplied by the reciprocal of the larger answer.

This gives a final value of $\dfrac{81}{100}$

What are the two consecutive, even numbers?   **[6 marks]**

5 A curve has equation $y = 3x^4 + 18x^3 - 15x^2 + 6$

a **i** Find $\dfrac{dy}{dx}$ [2 marks]

_____

_____

_____

_____

_____

_____

ii Find $\dfrac{d^2 y}{dx^2}$ [1 mark]

_____

_____

_____

_____

_____

_____

_____

b The point T, with coordinates (1, 12), lies on the curve.

Determine whether y is increasing or decreasing at point T.

Give a reason for your answer. [2 marks]

_____

_____

_____

_____

_____

_____

_____

_____

_____

_____

_____

5 c Find the coordinates of the three stationary points of y

In each case, determine the nature of the stationary point. **[8 marks]**

6 a On the same set of axes, given below, sketch the graphs of

    i $y = \log_{10} x$ and   **[2 marks]**

    ii $y = 10^x$ **[2 marks]**

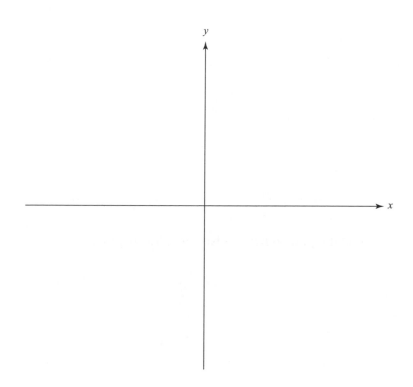

    iii What is the relationship between $y = \log_{10} x$ and $y = 10^x$?

          How is this relationship shown in their graphs?   **[2 marks]**

_____

_____

_____

_____

b Solve this equation. $10^{2x+5} = 4$

     Give your answer to 3 significant figures.   **[3 marks]**

_____

_____

_____

_____

_____

_____

7 a Three points A, B and C have coordinates (−4, 6), (4, 4) and (1, −2) respectively.

    i Find the equation of the perpendicular bisector of the line AB   **[4 marks]**

    ii Find the equation of the perpendicular bisector of the line BC   **[1 mark]**

    b Calculate the coordinates of the point of intersection of the lines AB and BC   **[4 marks]**

7 c Hence find the equation of the circumcircle of the triangle with vertices A, B and C.

Write your answer in the form $(x - a)^2 + (y - b)^2 = r^2$

where a, b and r^2 are rational numbers. **[4 marks]**

8 a Complete this binomial expansion which has been started for you.

$$(a + b)^8 = a^8 + 8a^7b + 28a^6b^2 + \ldots$$ **[2 marks]**

8   b   A multiple choice test consists of eight questions.

Each question has three answers to choose from, only one of which is correct.

Sidney tries to guess the answer to each of the eight questions.

What is the probability, to 3 significant figures, that Sidney guesses

   i   None of the answers correctly,   **[2 marks]**

_____

_____

_____

_____

_____

_____

_____

_____

_____

_____

   ii   Exactly two of the answers correctly?   **[2 marks]**

_____

_____

_____

_____

_____

_____

_____

_____

_____

_____

_____

_____

_____

_____

End of section A

Answer **all** questions in the spaces provided.

9   A bag contains a large number of objects of varying colours and shapes. These are listed in the table.

	Blue	Green	Red	Total
Circle	13	2	29	44
Square	16	6	6	28
Triangle	19	7	14	40
Total	48	15	49	112

a   Calculate the probability that a randomly-chosen shape from the bag

    i   Is a blue square, **[2 marks]**

    ii   Is a square given that it is blue. **[2 marks]**

b   Determine if the shape being red and the shape being a triangle are independent properties. Explain your reasoning. **[2 marks]**

9   c   Draw a Venn diagram to show the probabilities of the events of being red or not red and of being a triangle or not a triangle.   **[2 marks]**

10   For a binomial distribution $X \sim \text{Bin}(18, 0.4)$ calculate the following probabilities.

a $P(X = 3)$ **[1 mark]**

b $P(X \le 5)$ **[1 mark]**

c $P(X > 8)$ **[1 mark]**

11   A manufacturer wants to investigate the preferred percentage of cocoa powder to use in its chocolate bars. It decides to survey 120 people in its local town. It believes that different groups of people will have different preferences so makes sure that its sample of people has the proportions of each group specified in the table below.

Appearance	Office worker	Shop worker	Student	Retired	Other
% of population	25%	25%	10%	15%	25%
Number in sample	30	30	12	18	30

a   What is the name of this type of sampling method?   [1 mark]

_____

_____

_____

b   The following data is taken from the 12 students sampled showing their preferred percentage of cocoa in chocolate bars. Higher percentages are for darker chocolate.

1%   5%   9%   21%   22%   25%   31%   37%   38%   60%   63%   68%

Calculate the five-number summary and interquartile range for the data.   [4 marks]

_____

_____

_____

_____

_____

_____

_____

_____

_____

_____

_____

_____

_____

_____

_____

_____

_____

_____

_____

11 c The data for the office workers is collected and given in this box-and-whisker plot.

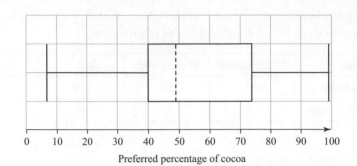

Preferred percentage of cocoa

Give three different comparisons between the preferred percentage of cocoa of the students and the office workers. **[3 marks]**

_____

_____

_____

_____

_____

_____

_____

_____

_____

_____

_____

_____

_____

_____

_____

_____

_____

_____

_____

_____

_____

_____

_____

_____

_____

12   People in a countryside town and in a coastal town are sampled to see if there are differences between the ages of the residents.

The results of the survey are presented in this cumulative frequency graph.

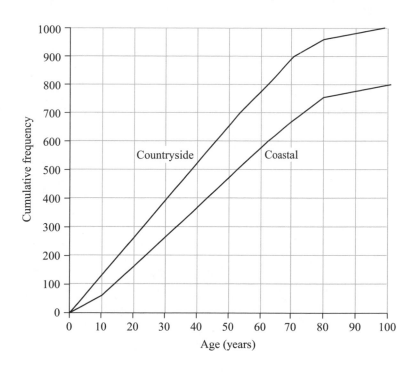

a   Estimate the percentage of people in each town who are 65 years or older.   **[3 marks]**

_____

_____

_____

_____

_____

_____

_____

_____

_____

b   The mayor of the coastal town claims that the average age of someone in their town is 43 years. Explain how they determined that number.   **[2 marks]**

_____

_____

_____

_____

_____

_____

12 c It is believed that political preference is associated with age.

Explain why cluster sampling by town should not be used to survey political preferences.

State which assumption in the use of cluster sampling is not present. **[3 marks]**

End of questions

Mathematical formulae
For A Level Maths

Pure Mathematics
Binomial series

$$(a+b)^n = a^n + \binom{n}{1}a^{n-1}b + \binom{n}{2}a^{n-2}b^2 + \cdots + \binom{n}{r}a^{n-r}b^r + \cdots + b^n \qquad (n \in \mathbb{N})$$

$$\text{where } \binom{n}{r} = {}^nC_r = \frac{n!}{r!(n-r)!}$$

$$(1+x)^n = 1 + nx + \frac{n(n-1)}{1.2}x^2 + \dots + \frac{n(n-1)\dots(n-r+1)}{1.2\dots r}x^r + \dots \qquad (|x|<1, \ n \in \mathbb{Q})$$

Arithmetic series

$$S_n = \frac{1}{2}n(a+l) = \frac{1}{2}n[2a+(n-1)d]$$

Geometric series

$$S_n = \frac{a(1-r^n)}{1-r} \qquad\qquad S_\infty = \frac{a}{1-r} \text{ for } |r|<1$$

Trigonometry: small angles

For small angle θ, measured in radians:

$$\sin\theta \approx \theta \qquad \cos\theta \approx 1 - \frac{\theta^2}{2} \qquad \tan\theta \approx \theta$$

Trigonometric identities

$$\sin(A \pm B) = \sin A \cos B \pm \cos A \sin B$$

$$\cos(A \pm B) = \cos A \cos B \mp \sin A \sin B$$

$$\tan(A \pm B) = \frac{\tan A \pm \tan B}{1 \mp \tan A \tan B} \qquad \left(A \pm B \neq \left(k+\frac{1}{2}\right)\pi\right)$$

Differentiation

$f(x)$	$f'(x)$
$\tan x$	$\sec^2 x$
$\operatorname{cosec} x$	$-\operatorname{cosec} x \cot x$
$\sec x$	$\sec x \tan x$
$\cot x$	$-\operatorname{cosec}^2 x$
$\dfrac{f(x)}{g(x)}$	$\dfrac{f'(x)g(x) - f(x)g'(x)}{(g(x))^2}$

Differentiation from first principles

$$f'(x) = \lim_{h \to 0} \frac{f(x+h) - f(x)}{h}$$

Integration

$$\int u \frac{dv}{dx} dx = uv - \int v \frac{du}{dx} dx$$

$$\frac{f'(x)}{f(x)} = \ln|f(x)| + c$$

$f(x)$	$\int f(x)dx$
$\tan x$	$\ln|\sec x| + c$
$\cot x$	$\ln|\sin x| + c$

Numerical solution of equations

The Newton–Raphson iteration for solving $f(x) = 0$: $x_{n+1} = x_n - \dfrac{f(x_n)}{f'(x_n)}$

Numerical integration

The trapezium rule: $\displaystyle\int_a^b y\,dx \approx \frac{1}{2}h\{(y_0 + y_n) + 2(y_1 + y_2 + \cdots + y_{n-1})\}$, where $h = \dfrac{b-a}{n}$

Mechanics

Constant acceleration

$$s = ut + \frac{1}{2}at^2 \qquad\qquad \mathbf{s} = \mathbf{u}t + \frac{1}{2}\mathbf{a}t^2$$

$$s = vt - \frac{1}{2}at^2 \qquad\qquad \mathbf{s} = \mathbf{v}t - \frac{1}{2}\mathbf{a}t^2$$

$$v = u + at \qquad\qquad \mathbf{v} = \mathbf{u} + \mathbf{a}t$$

$$s = \frac{1}{2}(u+v)t \qquad\qquad \mathbf{s} = \frac{1}{2}(\mathbf{u}+\mathbf{v})t$$

$$v^2 = u^2 + 2as$$

Probability and Statistics

Probability

$$P(A \cup B) = P(A) + P(B) - P(A \cap B)$$

$$P(A \cap B) = P(A) \times P(B|A)$$

Discrete distributions

Distribution of X	$P(X = x)$	Mean	Variance
Binomial B(n, p)	$\binom{n}{x} p^x (1-p)^x$	np	$np(1-p)$

Sampling distributions

For a random sample of n observations from N(μ, σ^2)

$$\frac{\bar{X} - \mu}{\frac{\sigma}{\sqrt{n}}} \sim \text{N}(0, 1)$$

Objectives checklist

Use these checklists to assess your confidence with the topics in AS Level Maths.

Ch	Objective	MyMaths	InvisiPen	No	Almost	Yes!
1 Algebra 1	Use direct proof, proof by exhaustion and counter-examples.	2252, 2253	01S1A	☐	☐	☐
	Use and manipulate index laws.	2033–2035	01S2B	☐	☐	☐
	Manipulate surds and rationalise a denominator.	2036, 2037	01S3A	☐	☐	☐
	Solve quadratic equations and sketch quadratic curves.	2014–2017, 2024–2026	01S4B	☐	☐	☐
	Understand and use coordinate geometry.	2002–2004, 2020, 2021	–	☐	☐	☐
	Understand and solve simultaneous equations.	2005, 2018	01S5A, 01S6B	☐	☐	☐
	Understand and solve inequalities.	2008, 2009	01S7A	☐	☐	☐
2 Polynomials and the binomial theorem	Manipulate, simplify and factorise polynomials.	2006	02S1A	☐	☐	☐
	Understand, explore and use the binomial theorem.	2041	02S2B	☐	☐	☐
	Divide polynomials by algebraic expressions.	2043	–	☐	☐	☐
	Understand and use the factor theorem.	2042	02S3A	☐	☐	☐
	Use a variety of techniques to analyse a function and sketch its graph.	2022–2024, 2027, 2258	02S4B	☐	☐	☐
3 Trigonometry	Calculate the values of sine, cosine and tangent for angles of any size.	2047, 2048, 2257	–	☐	☐	☐
	Use the two identities $\sin^2\theta + \cos^2\theta = 1$ and $\tan\theta = \sin\theta \div \cos\theta$ and recognise $x^2 + y^2 = r^2$ as the equation of a circle.	2053, 2257, 2284	03S1A	☐	☐	☐
	Sketch graphs of trigonometric functions and describe their main features.	2047, 2048	03S1A	☐	☐	☐
	Solve various types of trigonometric equations.	2047, 2048, 2053, 2257, 2284, 2285	03S1A	☐	☐	☐
	Use the sine and cosine rules and the area formula for a triangle.	2045, 2046	03S2B	☐	☐	☐
4 Differentiation and integration	Differentiate from first principles.	2028	04S1A	☐	☐	☐
	Differentiate functions composed of terms of the form ax^n	2029	04S2B	☐	☐	☐
	Use differentiation to calculate rates of change.	2269, 2270	–	☐	☐	☐
	Work out equations, tangents and normals.	2030	04S4B	☐	☐	☐
	Work out turning points and determine their nature.	2270	04S5A	☐	☐	☐
	Work out and interpret the second derivative.	2270	04S5A	☐	☐	☐
	Work out the integral of a function.	2054, 2055	04S6B	☐	☐	☐
	Understand and calculate definite integrals. Use them to calculate the area under a curve.	2056, 2273	04S7A	☐	☐	☐

Objectives checklist

Ch	Objective	MyMaths	InvisiPen	No	Almost	Yes!
5 Exponentials and logarithms	Convert between powers and logarithms.	2062	05S1B	☐	☐	☐
	Manipulate expressions and solve equations involving powers and logarithms.	2062, 2063, 2257	05S1B, 05S2A	☐	☐	☐
	Use the exponential functions $y = a^x$, $y = e^x$, $y = e^{kx}$ and their graphs.	2061, 2133, 2134, 2136	05S2A, 05S3B	☐	☐	☐
	Verify and use mathematical models, including those of the form $y = ax^n$ and $y = kb^x$	2268	05S4A	☐	☐	☐
	Consider limitations of exponential models.	–	–	☐	☐	☐
6 Vectors	Identify vector quantities and scalar quantities.	2206	–	☐	☐	☐
	Solve geometric problems in two dimensions using vector methods.	2206	06S1B	☐	☐	☐
	Solve problems involving displacements, velocities and forces.	2206	–	☐	☐	☐
	Find and use the components of a vector.	2207	06S2A	☐	☐	☐
	Find the magnitude and direction of a vector expressed in component form.	2207	06S2A	☐	☐	☐
	Use position vectors to find displacements and distances.	2207	–	☐	☐	☐
7 Units and kinematics	Understand and use standard SI units and convert between them and other metric units.	2183	–	☐	☐	☐
	Calculate average speed and average velocity.	2183	07S1B	☐	☐	☐
	Draw and interpret graphs of displacement and velocity against time.	2183	07S2A	☐	☐	☐
	Derive and use the formulae for motion in a straight line with constant acceleration.	2184	07S3B	☐	☐	☐
	Use calculus to solve problems involving variable acceleration.	2289	07S4A	☐	☐	☐
8 Forces and Newton's laws	Resolve in two perpendicular directions for a particle in equilibrium.	2186, 2293	08S1A	☐	☐	☐
	Calculate the magnitude and direction of the resultant force acting on a particle.	2293	–	☐	☐	☐
	Resolve for a particle moving with constant acceleration. Work out acceleration of forces.	2187, 2293	08S2B	☐	☐	☐
	Understand the connection between the mass and the weight of an object. Know that weight changes depending on where the object is.	2185, 2187	08S3A	☐	☐	☐
	Resolve for "connected objects", such as an object in a lift.	2188	–	☐	☐	☐
	Resolve for particles moving with constant acceleration connected by string over pulleys.	2188	08S4B	☐	☐	☐

Ch	Objective	MyMaths	InvisiPen	No	Almost	Yes!
9 Collecting, representing and interpreting data	Distinguish a population and its parameters from a sample and its statistics.	2275	09S1A	☐	☐	☐
	Identify and name sampling methods.	2275	09S1A	☐	☐	☐
	Highlight sources of bias in a sampling method.	2275	–	☐	☐	☐
	Read continuous data given in box-and-whisper plots, histograms and cumulative frequency diagrams.	2276–2278	09S3A	☐	☐	☐
	Plot scatter diagrams and use them to identify types and strength of correlation.	2283	09S4B	☐	☐	☐
	Use scatter diagrams and rules using quantities to identify outliers.	2283	09S4B	☐	☐	☐
	Summarise raw data using appropriate measures of location and spread.	2279 –2282	09S2B	☐	☐	☐
10 Probability and discrete random variables	Use the vocabulary of probability theory, including the following terms: random experiment, sample space, independent events and mutually exclusive events.	2093	10S1B	☐	☐	☐
	Solve the problems involving mutually exclusive and independent events using the addition and multiplication rules.	2093, 2094	10S1B	☐	☐	☐
	Use a probability function or a given context to find the probability distribution and probabilities for particular events.	2114	–	☐	☐	☐
	Recognise and solve problems relating to experiments which can be modelled by the binomial distribution.	2110, 2111	10S2A	☐	☐	☐
11 Hypothesis testing 1	Understand the terms null hypothesis and alternative hypothesis.	2115	11S1A, 11S2B	☐	☐	☐
	Understand the terms critical value, critical region and significance level.	2115	11S1A	☐	☐	☐
	Calculate the critical region.	2115	11S1A	☐	☐	☐
	Calculate the p-value.	2115	–	☐	☐	☐
	Decide whether to reject or accept the null hypothesis.	2115	11S1A	☐	☐	☐
	Make a conclusion based on whether you reject or accept the null hypothesis.	2115	11S2B	☐	☐	☐

Answers

Paper 1 (Set A)

1 $x < -4$, $x > 4$

2 a i $\dfrac{x^4}{4} + c$

ii $-\dfrac{1}{x} + c$ or $-x^{-1} + c$

iii $-\dfrac{1}{4x^4} + c$ or $-\dfrac{1}{4}x^{-4} + c$

b $-3x^{-1} - \dfrac{1}{2}x^4 - \dfrac{3}{4}x^{-4} - 2x + c$ or

$-3\dfrac{1}{x} - \dfrac{1}{2}x^4 - \dfrac{3}{4}\dfrac{1}{x^4} - 2x + c$

3 a $f(6) = 6^4 - 6^3 - 22(6)^2 - 44(6) - 24 = 0$
$\therefore x - 6$ is a factor
$f(-1) = (-1)^4 - (-1)^3 - 22(-1)^2 - 44(-1)$
$-24 = 0$
$\therefore x + 1$ is a factor

b $f(x) = (x - 6)(x + 1)(x + 2)^2$

4 $1 : \dfrac{11}{5}$ or $1 : 2.2$

5 $x = -1 + \dfrac{\sqrt{22}}{2}$ or $x = \dfrac{-2 + \sqrt{22}}{2}$

6 a Radius = 2, Centre = (3, 6)

b Translation $\begin{pmatrix} 3 \\ 6 \end{pmatrix}$

c i $y = -x + 9$
ii $(3 + \sqrt{2}, 6 - \sqrt{2})$ and $(3 - \sqrt{2}, 6 + \sqrt{2})$

7 a $\theta = 0°, 90°$
b $\theta = 9°, 27°, 45°, 63°, 81°$

8 $26a^3 + \dfrac{2}{a}$

9 37.6 (3 sf)

10 170 m

11 a $y = \pm 12$ N
b $\pm 18.9°$

12 a 63 m
b 36 m s⁻¹
c 2.71 s
d Air resistance will reduce the particle's acceleration. The time taken will increase.

13 a Newton's second law, applied vertically
$8g - T = 8a$
$80 - T = 8a$

b $a = 2\dfrac{4}{13}$ m s⁻² = 2.31 m s⁻² (3 sf)

$T = 61\dfrac{7}{13}$ N = 61.5 N (3 sf)

Paper 2 (Set A)

1 $\dfrac{5 + 2\sqrt{3}}{13}$

2 0

3 18 and 33

4 $\log_{10}\sqrt{\dfrac{b}{c}}$

5 $-\dfrac{7}{8}$

6 $q'(x) = \lim_{h \to 0} \dfrac{q(x + h) - q(x)}{h}$

$q'(x) = \lim_{h \to 0} \dfrac{4(x + h)^2 - 2(x + h) - (4x^2 - 2x)}{h}$

$q'(x) = \lim_{h \to 0} \dfrac{4x^2 + 8xh + 4h^2 - 2x - 2h - 4x^2 + 2x}{h}$

$q'(x) = \lim_{h \to 0} (8x + 4h - 2)$

$q'(x) = 8x - 2$

7 a i $\log_{10} y = \log_{10}(ab^x)$
$\log_{10} y = \log_{10} a + \log_{10} b^x$
$\log_{10} y = \log_{10} a + x \log_{10} b$
$\log_{10} y = x \log_{10} b + \log_{10} a$
ii Points lie on straight line
$\Rightarrow$ Assertion correct
b $a = 1.6$, $b = 1.1$

8 a i $y = \dfrac{3}{2}x + \dfrac{1}{2}$
ii Intersection of line and either circle

$x^2 + \left(\dfrac{3}{2}x + \dfrac{1}{2}\right)^2 - 2x + 3\left(\dfrac{3}{2}x + \dfrac{1}{2}\right) = 9$

$x^2 + \dfrac{9}{4}x^2 + \dfrac{3}{2}x + \dfrac{1}{4} - 2x + \dfrac{9}{2}x + \dfrac{3}{2} = 9$

or $4(x + 2)^2 + 4\left(\dfrac{3}{2}x\right)^2 = 45$

$4x^2 + 16x + 16 + 9x^2 = 45$
$13x^2 + 16x - 29 = 0$
$(13x + 29)(x - 1) = 0$

$x = 1, \quad -\dfrac{29}{13}$

$y = \dfrac{3}{2} \times 1 + \dfrac{1}{2} = 2$ or

$y = \dfrac{3}{2} \times -\dfrac{29}{13} + \dfrac{1}{2} = -\dfrac{37}{13}$

Coordinates are A $(1, 2)$ and B $\left(-\dfrac{29}{13}, -\dfrac{37}{13}\right)$

b 5.824

9 a $h = 4 - 7x$
b $2(3x \times h + 4x \times h + 3x \times 4x)$
$= 14hx + 24x^2$
$= 14(4 - 7x) + 24x^2$
$= 56x - 74x^2$

c i $10\dfrac{22}{37}$ cm²

ii $1\dfrac{13}{37}$ cm by $1\dfrac{5}{37}$ cm by $1\dfrac{19}{37}$ cm

10 a Mutually exclusive
b 0.36

11 a Cluster sampling
b $\dfrac{3}{11}$ or 0.273 (3 sf)
c Knowing the amount of meat eaten by a person tells you nothing about the amount of cheese eaten by a person, and vice versa.

12 a,b

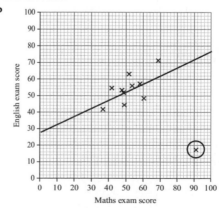

The point is very different from the rest of the class.

c 57 – 68 marks

d

$0 \leq x < 40$	1	
$40 \leq x < 45$	1	
$45 \leq x < 50$	3	
$50 \leq x < 55$	2	
$55 \leq x < 60$	1	
$60 \leq x < 70$	2	
$70 \leq x < 100$	1	

e i $\dfrac{6}{11}$ or 0.545 (3 sf)

    ii $\dfrac{1}{2}$ or 0.5

13 a i Each packet is given a number from 1 to 3000…

      …and 18 different random numbers are generated in this range.

      Use the packets corresponding to those numbers.

    ii The sample draws at random from the population and any packet is equally likely to be chosen.

  b Let p = the probability that a packet is not fit for sale.

    H_0: $p = 0.06$

    H_1: $p > 0.06$

    Let X = the number of packet that are unfit for sale.

    $X \sim$ Bin (18, 0.06)

    p-value = $P(X \geq 3) = 0.0898$ (3 sf)

    or

    critical region $X \geq 4$

    $0.0898 > 0.05$

    or

    $3 < 4$

    There is insufficient evidence to reject the null hypothesis. The probability that a packet is not fit for sale has not increased

Paper 1 (Set B)

1 e^{2x} and $(e^x)^2$

2 $180°$

3 a 1.4314

  b 2.4771

  c 1.0458

4 a

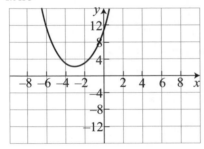

  b

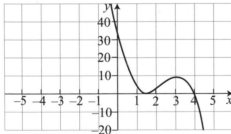

  c

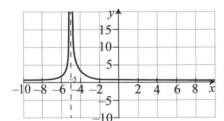

  d

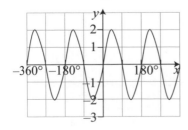

5 $t^2 - t = t(t - 1)$

  t even $\Rightarrow t - 1$ is odd

  t odd $\Rightarrow t - 1$ is even

  even $\times$ odd = even

  odd $\times$ even = even

  $\therefore t^2 - t$ is even

  OR

  t even $\Rightarrow t^2$ even

  even − even = even

  t odd $\Rightarrow t^2$ odd

  odd − odd = even

  $\therefore t^2 - t$ is even

6 a $6x^2 + 8x - 4 = p(2x - 1)$

    $6x^2 + (8 - 2p)x - 4 + p = 0$

    $6x^2 + 2(4 - p)x + p - 4 = 0$

  b i Discriminant > 0

    $(8 - 2p)^2 - 4 \times 6 \times (p - 4) > 0$

    $64 - 32p + 4p^2 - 24p + 96 > 0$

    $4p^2 - 56p + 160 > 0$

    $p^2 - 14p + 40 > 0$

    ii $p < 4$ or $p > 10$

7 37.6 cm²

8 a $256 - 1024x + 1792x^2 + \dots$

  b 245.9

9 a 1.61 hrs

  b i The population grows without limit at large times.

    ii The growth in the population should slow down at larger times.

10 $\dfrac{dy}{dx} = 12x^2 - 12x + 6$

  $= 12\left(x^2 - x + \dfrac{1}{2} \right)$

  $= 12\left(\left(x - \dfrac{1}{2} \right)^2 - \dfrac{1}{4} + \dfrac{1}{2} \right)$

  $= 12\left(\left(x - \dfrac{1}{2} \right)^2 + \dfrac{1}{4} \right)$

  > 0 (Any number squared is positive)

  A positive derivative indicates an increasing function

11 kg m s⁻²

12 $\overrightarrow{AB} = \overrightarrow{OB} - \overrightarrow{OA} = 2i - 3j$

  $\overrightarrow{BC} = \overrightarrow{OC} - \overrightarrow{OB} = 4i - 6j$

  $\overrightarrow{AC} = \overrightarrow{OC} - \overrightarrow{OA} = 6i - 9j$

  $\overrightarrow{BC} = 2\overrightarrow{AB}$ or $\overrightarrow{AC} = 3\overrightarrow{AB}$ or

  $\overrightarrow{AC} = \dfrac{3}{2}\overrightarrow{BC}$

  $\overrightarrow{AB}$ and $\overrightarrow{BC}$ are parallel and both pass through point B   or

  $\overrightarrow{AB}$ and $\overrightarrow{AC}$ are parallel and both pass through point A   or

  $\overrightarrow{AC}$ and $\overrightarrow{BC}$ are parallel and both pass through point C

  $\therefore$ A, B and C are collinear.

13 a Stationary

  b 7.5 m s⁻¹

14 a $12 - 6t$ m s⁻²

  b $s = 6t^2 - t^3 + c$

    $c = 5$

    $s = 6 \times 4^2 - 4^3 + 5$

    $= 96 - 64 + 5$

    $= 37$

  c 0 and 4 s

15 a 0.15 m s⁻²

  b 280 N

Paper 2 (Set B)

1 $\dfrac{1}{x^2}$

2 $\dfrac{2}{3}$

3 Linda
Right-hand side must be zero before factorising
OR
Correct answers are 1 and 5

4 $x = -1, y = 3$ or $x = 9, y = -\dfrac{19}{2}$

5 a $\theta = 57.5°, 122.5°, 216.4°, 323.6°$
  b $x = 13.8°, 43.3°, 93.2°, 146.8°$

6 a £750
  b £458.05
  c 2.94 years (3 sf)
  d
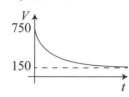

  e Not a realistic model, as the model suggests that the value of the phone will never be less than £150

7 $(2-\sqrt{3})$ cm

8 a $y = -2x^2 + x + 1$
  b $-\dfrac{1}{2}$
  c $\dfrac{9}{8}$

9 a $\dfrac{du}{dx} = 15x^2 + 8x$
  b $u = 5x^3 + 4x^2 - 1$

10 a 5%
  b $X = 37$

11 a 0.0075
  b i 0.255
  ii 0.027 (3 sf)

12 a i Take data from any members of the population that are available.
  ii 4, 1.31 (3 sf)
  b i Not every member of the population is as likely to have an equal chance of being selected using opportunity sampling.
  ii The first sample has a higher average.
    The first sample is more spread out around that average.

13 a Frequencies: 12, 7, 4, 4, 3, 0
  b $\dfrac{1}{6}$ or 0.167
  c i Any two of
    Fixed number of trials; there are 14 days over the two weeks.
    Fixed probability of success; it is $\dfrac{1}{6}$ (FT answer to part **b**).
    There are two possible outcomes; ≥ 6 litres sold or < 6 litres sold.
    Trials are independent; assuming the store has stock to sell then the amount people buy on one day shouldn't affect the amount sold on any other day.
  ii Let p = the probability that 6 kg or more of ice cream are sold on any one day.
    $H_0: p = \dfrac{1}{6}$, $H_1: p < \dfrac{1}{6}$
    Let X = the number of days when less than 6 kg of ice cream are sold.
    $X \sim B\left(14, \dfrac{1}{6}\right)$
    p-value = $P(X = 0) = 0.0779$
    $0.0779 < 10\%$
    There is sufficient evidence to reject the null hypothesis.
    Less ice cream was sold during the rainy weeks.

Paper 1 (Set C)

1 a $L \propto \dfrac{1}{W}$ and $L = \dfrac{k}{W}$
  b C

2 a
$115 = 5 \times 23$ $\Rightarrow$ not prime
$116 = 2^2 \times 29$ $\Rightarrow$ not prime
$117 = 3 \times 39$ $\Rightarrow$ not prime
$118 = 2 \times 59$ $\Rightarrow$ not prime
$119 = 7 \times 17$ $\Rightarrow$ not prime
$120 = 2^3 \times 3 \times 5$ $\Rightarrow$ not prime
$121 = 11^2$ $\Rightarrow$ not prime
$122 = 2 \times 61$ $\Rightarrow$ not prime
$123 = 3 \times 41$ $\Rightarrow$ not prime
$124 = 2^2 \times 31$ $\Rightarrow$ not prime
$125 = 5^3$ $\Rightarrow$ not prime
All numbers have factors (other than 1 and the number itself)
∴ The statement is true.
There are no primes between 115 and 125

  b For example, $a = \sqrt{2}, b = \sqrt{2}$

3 a 135°
  b i $-\dfrac{8}{17}$
  ii $-\dfrac{8}{15}$
  c $\left(\dfrac{\cos x}{1 - \sin x}\right) \cdot \left(\dfrac{1 + \sin x}{1 + \sin x}\right)$
    $= \dfrac{\cos x + \cos x \sin x}{1 - \sin^2 x}$
    $= \dfrac{\cos x + \cos x \sin x}{\cos^2 x}$
    $= \dfrac{1 + \sin x}{\cos x}$
    OR
    $\left(\dfrac{\cos x}{1 - \sin x}\right) \cdot \dfrac{\cos x}{\cos x}$
    $= \dfrac{1 - \sin^2 x}{(1 - \sin x)\cos x}$
    $= \dfrac{(1 - \sin x)(1 + \sin x)}{(1 - \sin x)\cos x}$
    $= \dfrac{1 + \sin x}{\cos x}$

4 a i $-8 < x \le \dfrac{16}{3}$
  ii $x < \dfrac{17}{10}$
  b The workings assume $x + 4 > 0$ – as the inequality is not reversed. However this is not true if $x < -4$
    OR
    Correct solution is $x < -6$ or $x > -4$
    ∴ Habib's solution is not correct.

5 a i $\log_{10} y = \log_{10} (kx^n)$
    $\log_{10} y = \log_{10} k + \log_{10} x^n$
    $\log_{10} y = \log_{10} k + n\log_{10} x$
  ii Points lie on a straight line
    $\Rightarrow$ Belief correct
  b $n = 1.6, k = 2.5$

6 a
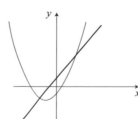

  b
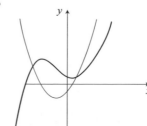

7 a 28.3
b i $\mathbf{d}+\mathbf{e}+\mathbf{f}$
   ii $\overrightarrow{OD}+\overrightarrow{OE}+\overrightarrow{OF}$

8 41.4°

9 $2.87\,\mathrm{m\,s^{-2}}$ (3 sf)

10 a $53.3\,\mathrm{m\,s^{-1}}$ (3 sf)
b The acceleration is *not* constant.
c Weight is acting downwards.
   Air resistance is acting upwards.
   The forces are equal and opposite.
   785 N (3 sf)

11 a $2.67\,\mathrm{m\,s^{-2}}$ (3 sf)
b $0\,\mathrm{s}$ and $25\,\mathrm{s}$
c $69\,\mathrm{s}$

12 a $v=3t^2-4t+3$
b $s=t^3-2t^2+3t+c$
   $0=0+c$
   $s=t^3-2t^2+3t$
   $0=t^3-2t^2+3t$
    $=t(t^2-2t+3)$
   'b^2-4ac' $=2^2-4\times1\times3$
      $=-8$
   $-8<0\Rightarrow$ no real solutions
   Therefore only solution at $t=0$

Paper 2 (Set C)

1 $2\sqrt{2}$

2 2

3 a He has divided through by $\sin x$ and therefore lost the solutions
   to $\sin x=0$
   OR
   Missed out $x=0°$, $x=180°$
b $x=0°,\ 19.5°,\ 160.5°,\ 180°$

4 18 and 20

5 a i $\dfrac{dy}{dx}=12x^3+54x^2-30x$

   ii $\dfrac{d^2y}{dx^2}=36x^2+108x-30$

b When $x=1$, $\dfrac{dy}{dx}=36$

   $\dfrac{dy}{dx}>0\Rightarrow$ increasing

c $(-5,-744)$ is a minimum, $(0,6)$ is a maximum, $\left(\dfrac{1}{2},\dfrac{75}{16}\right)$ is a minimum

6 a i,ii

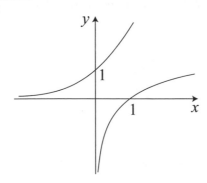

   iii They are inverses of one another.
     Their graphs are reflections of one another in the line $y=x$
b $x=-2.20$

7 a i $y=4x+5$

   ii $y=-\dfrac{1}{2}x+\dfrac{9}{4}$

b $\left(-\dfrac{11}{18},\dfrac{23}{9}\right)$

c $\left(x+\dfrac{11}{18}\right)^2+\left(y-\dfrac{23}{9}\right)^2=\dfrac{7565}{324}$

8 a $\ldots+56a^5b^3+70a^4b^4+56a^3b^5+28a^2b^6+8ab^7+b^8$
b i 0.0390
   ii 0.273

9 a i $\dfrac{1}{7}$

   ii $\dfrac{1}{3}$

b P(red and triangle) $=\dfrac{14}{112}=\dfrac{1}{8}$

   P(red) $\times$ P(triangle) $=\dfrac{49}{112}\times\dfrac{40}{112}=\dfrac{5}{32}$
   $\neq$ P(red and triangle)
   $\Rightarrow$ The events are not independent.

c

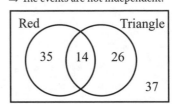

10 a 0.0246
b 0.209
c 0.263

11 a Stratified
b Min = 1, LQ = 15, Med = 28, UQ = 49, Max = 68
   IQR = 34
c For example
   The office workers liked darker chocolate on average.
   Both groups have the same interquartile range.
   The students have a smaller total range, indicating more
   of a consensus of opinion.

12 a Countryside $\approx$ 17%, Coastal $\approx$ 24%
b They have found the median age.
   Using a line across from 400 people, 50% of 800, and down to
   43 years.
c Cluster sampling is used when each cluster is similar to any
   other.
   The political preferences in the two towns are unlikely to be
   similar...
   ...because the ages aren't similar.